Medical Oncology Series –
Epithelial Ovarian Cancer

© 2014 I H Zubairi. All rights reserved.
ISBN 978-1-312-76142-1

Anyone seeking to apply or consult this text is expected to use independent medical judgement in the context of individual clinical circumstances to determine any patient's care or treatment. The use of this text is at your own risk.

Epithelial Ovarian Cancer

Contents

History 7
Risk factors 7
Symptoms of ovarian cancer 8
Protective factors 9
Favourable prognostic factors 9
Investigations 10
Risk of Malignancy Index (RMI) 12
Histology and response to chemotherapy .. 13
Staging of Ovarian Cancer 14
Treatment – Surgery 16
Treatment - Chemotherapy 17
Survival 19
Follow up 20
Palliative Care 21
CA125 Relapse 23
Adjuvant chemotherapy 24
First line chemotherapy for epithelial ovarian cancer 25
Second line chemotherapy for platinum sensitive epithelial ovarian cancer 27
Second line chemotherapy for platinum resistant ovarian cancer 28
AURELIA study chemo +/- bevacizumab in platinum resistance ovarian cancer 29
Few of Current Clinical Trials in UK 30

History

- Age
- Fertility
- Pelvic pain and pressure
- Discharge / bleeding
- Fever (infection)
- Family history

Risk factors

- Age
- Obesity
- Hormone replacement therapy
- Family history of ovarian cancer
- BRCA I mutation 35-46% probability of developing cancer
- BRCA II mutation 13-23% probability of developing cancer
- Hereditary nonpolyposis colon cancer
- Infertility
- Endometriosis
- Cigarette smoking ↑ risk for mucinous

Symptoms of ovarian cancer

- Pelvic pain
- Abdominal pain
- Urinary urgency
- Urinary frequency
- Increased abdominal size/bloating
- Difficulty eating
- Sensation of fullness

- Cramping
- Fear of recurrence / disease progression
- Indigestion
- Sexual dysfunction
- Vomiting
- Weight gain or loss

Protective factors

- Oral contraceptives
- Breast feeding
- Tubal ligation
- Salpingo-oophorectomy

Favourable prognostic factors

- Young age
- Good performance status
- Earlier stage of cancer
- Lower grade tumour for stage I cancer
- Histology other than mucinous and clear cell
- Well-differentiated tumour.
- Small disease bulk prior to surgery
- No ascites.
- Smaller residual tumour following primary cytoreductive surgery.

Investigations

- Full blood count
- Biochemistry
- CA 125
- Ultrasound abdomen
- CT scan thorax, abdomen and pelvis
- Bowel studies if gastrointestinal cancer is being considered as a differential diagnosis/ mucinous histology
- Tissue biopsy
- Genetic studies

BRCA testing recommended in

1. Breast cancer diagnosed before age 50 years
2. Bilateral breast cancer
3. Both breast and ovarian cancers
4. Multiple breast cancers
5. Two or more primary types of *BRCA1*- or *BRCA2*-related cancers in a single family member
6. Male breast cancer
7. Ashkenazi Jewish ethnicity

Risk of Malignancy Index (RMI)

RMI is a scoring system that assesses risk of an ovarian mass being malignant

RMI = ultrasound score x menopausal score x CA-125 level in U/ml.

Feature	RMI 1	RMI 2
Ultrasound abnormalities: • Multilocular cyst • Solid areas • Bilateral lesions • Ascites • Intra-abdominal metastases	0 = no abnormality 1 = one abnormality 3 = two or more abnormalities	0 = none 1 = one abnormality 4 = two or more abnormalities
Menopausal score	1 = premenopausal 3 = postmenopausal	1 = premenopausal 4 = postmenopausal
CA-125	Quantity in U/ml	Quantity in U/ml

RMI 2 of over 200
- Sensitivity = 74 to 80%
- Specificity =- 89 to 92%
- Positive predictive value of around 80% of ovarian cancer. RMI 2 is more sensitive than RMI 1

Histology and response to chemotherapy

	Frequency	Response to 1st line chemo
Papillary Serous	70%	70-80%
Endometeriod	10%	70-80%
Clear Cell	10%	20-30%
Mucinous	5%	5%

Usual biopsy characteristics
– poorly differentiated adenocarcinoma CK7+, WT1+, PAX8+, CK20-.

Staging of Ovarian Cancer

Stage	
I	Growth limited to the ovaries.
IA	Growth limited to one ovary; no ascites present containing malignant cells. No tumour on the external surface; capsule intact.
IB	Growth limited to both ovaries; no ascites present containing malignant cells. No tumor on the external surfaces; capsules intact.
IC	Tumor either stage IA or IB, but with IC1 Surgical spill IC2 Capsule rupture before surgery or tumour on ovarian surface IC3 Malignant cells in ascites or with positive peritoneal washings
II	Growth involving one or both ovaries with pelvic extension.
IIA	Extension and/or metastases to the uterus and/or tubes.
IIB	Extension to other pelvic tissues.
III	Tumour involving one or both ovaries with histologically confirmed peritoneal implants outside the pelvis and/or positive regional lymph nodes. Superficial liver metastases equals stage III. Tumour is limited to the true

Stage	
	pelvis, but with histologically proven malignant extension to small bowel or omentum.
IIIA	III A Positive retroperitoneal lymph nodes and/or microscopic metastases beyond pelvis. IIIA1 Positive retroperitoneal lymph nodes only IIIA1(i) Metastases ≤ 10 mm IIIAI(ii) Metastases > 10 mm IIIA2 Microscpic ,extrapelvic, peritoneal metastases ≤ 2cm ± positive retroperitoneal lymph nodes
IIIB	Macroscopic, extrapelvic, peritoneal metastases ≤ 2 cm ± positive retroperitoneal lymph nodes. Includes extension to capsule of liver/spleen
IIIC	Macroscopic, extrapelvic, peritoneal metastases > 2 cm ± positive retroperitoneal lymph nodes. Includes extension to capsule of liver/spleen
IV	IVA Pleural effusion with positive cytology IVB Hepatic and/or splenic parenchymal metastases, metastases to extra-abdominal organs, including inguinal lymph nodes outside of abdominal cavity.

Treatment – Surgery

- Laparotomy, total abdominal hysterectomy, bilateral salpingo-ophorectomy (TAH, BSO) and omentectomy. Para aortic lymph node dissection.
- Unilateral salpingo-ophorectomy (USO) in selected patients with stage I disease requiring fertility preservation.
- For stage II and above – Optimal cytoreduction should be achieved (residual disease < 1 cm)
- Appendicectomy for mucinous cancer
- Consider surgery following neoadjuvant chemotherapy in patients who were initially unsuitable for treatment

- Aim for optimal debulking surgery defined as maximum diameter of the largest residual tumour nodule present within the peritoneal cavity at the completion of surgery < 1 cm.

Treatment - Chemotherapy

Stage IA – Chemo if high grade or grade 3. No consensus on role of chemotherapy for stage IA clear cell

Stage IB - Chemo if high grade or grade 3. No consensus on role of chemotherapy for stage IB clear cell

Stage IC- Chemo if high grade, all clear cell. Not for grade 2 endometrioid/ mucinous

Stage II – all patients for chemotherapy if performance status allows.

Chemotherapy stage I and II
- 6 x Carboplatin
- 3-6 x carboplatin and paclitaxel
- 6 x carboplatin and paclitaxel for clear cell histology

6 x Carboplatin AUC 5 IV over 1h on day 1; every 21 days

3-6 x Paclitaxel 175 mg/m^2 IV over 3 h and carboplatin AUC 5 IV over 30 min on day 1; every 21 days

Stage III and IV
- Debulking ----- adjuvant chemotherapy Paclitaxel 175 mg/m^2 IV over 3 h and carboplatin AUC 5 IV over 30 min on day 1; every 21 d for three to six cycles
- Primary chemotherapy x 3------ CT scan ------ discuss at MDT ------ interval surgery ----- 3 further cycles of chemotherapy

GFR calculation
Wright method use carboplatin AUC 5
Cockroft and Gault method use carboplatin AUC 6

Survival

Stage	5 year Survival	5 year DFS
I	80-90 %	70-85%
II	65-80 %	55-65%
IIIA	50 %	45 %
IIIB	40 %	25%
IIIC	30 %	20 %
IV	15 %	10 %

Follow up

Stage I and II	For first 2 years	Every 4 months
	Years 3- 5	Every 6 months
	Discharge	
Stage III and IV	For first 2 years	Every 3 months
	Years 3 and 4	Every 4 months
	Year 5	Every 6 months
	Years 6-8	Annually
Relapsed disease post chemotherapy	Asymptomatic	Every 4-6 weeks

Palliative Care

Intestinal obstruction:
- Abdominal x ray
- Nil by mouth
- IV fluids
- Morphine for pain control
- Dexamethasone 8mg/day iv/sc for 4-7 days
- Cyclizine 150 mg/day via 24 hour subcutaneous syringe driver. Reduces peristalsis
- Hyoscine butyl bromide 40-120 mg via 24 hour subcutaneous syringe driver. Reduces peristalsis
- Haloperidol 2 mg via 24 hour subcutaneous syringe driver. Can also be used as a stat dose. Prolongs QT interval.
- Levomepromazine 5-25 mg via 24 hour subcutaneous syringe driver. Prolongs QT interval. Hypotension.
- Naso gatric tube if no benefit with above
- Consider CT scan.
- Surgical opinion if single point of transition
- Palliative Medicine opinion
- Octreotide dose is 250-500 micrograms via 24 hour subcutaneous syringe driver.

- Avoid laxatives in complete obstruction. In the case of constipation Macrogol (Laxido) or docusate sodium can be used.
- Prokinetics – metoclopromide 30 mg via 24 hour subcutaneous syringe driver may help while the obstruction is resolving and the patient has no cramps. Not to be used in complete obstruction or cramps.
- Venting gastrostomy is rarely used.
- Total parenteral nutrition is rarely used.
- Bowel stenting

CA125 Relapse

In case previously elevated CA 125 had normalised and where CA-125 was not previously raised:

- Rise in CA125 ≥ 2 times upper limit of normal on 2 occasions not less than 1 week apart

Where previously elevated CA 125 had not normalised:

- Rise in CA125 ≥ 2 times nadir level on two occasions.

Normal CA125 premenopausal <35, postmenopausal <25 kU/L

Adjuvant chemotherapy

Evidence for benefit of adjuvant chemotherapy for ovarian cancer comes from two clinical trials

- International Collaboration for Ovarian Neoplasia (ICON) 1
- EORTC aCTION (adjuvant Chemotherapy in Ovarian Neoplasm) trial

Five year overall survival in favour of chemotherapy (82% versus 74% $P = 0.008$)

Refs:
- Colombo N, Guthrie D, Chiari S, et al. International Collaborative Ovarian Neoplasm trial 1: a randomized trial of adjuvant chemotherapy in women with early-stage ovarian cancer. *J Natl Cancer Inst* 2003;95:125-132.
- Trimbos JB, Vergote I, Bolis G, et al. Impact of adjuvant chemotherapy and surgical staging in early-stage ovarian carcinoma: European organisation for research and treatment of cancer-adjuvant chemotherapy in ovarian neoplasm trial. *J Natl Cancer Inst* 2003;95:113-125.
- Trimbos JB, Parmar M, Vergote I, et al. J Natl Cancer Inst. *Vol.* 95. 2003. International collaborative ovarian neoplasm trial 1 and adjuvant chemotherapy in ovarian neoplasm trial: two parallel randomized phase III trials of adjuvant chemotherapy in patients with early-stage ovarian carcinoma; *p.* 105-112

First line chemotherapy for epithelial ovarian cancer

Treatment	Improvement in OS/PFS
Carboplatin based chemotherapy	7-8% at 5 year follow up
MITO-7 carboplatin (AUC 2 mg/mL per min) plus paclitaxel (60 mg/m2) every week for 18 weeks VS 3 weekly carbo/taxol	Median PFS 17.3 vs 18.3 (p=0.66, HR 0.96) at 22.3 month follow up.
JGOG 3016 Stage II-1V ovarian cancer Carboplatin AUC 6 and paclitaxel 180 mg/m2 q 3 weeks vs. paclitaxel 80 mg/m2 d1,8,15 Primary endpoint = PFS	Median PFS 17.2 vs 28 months HR 0.71 in favour of DD regimen OS at 3 years 65.1 % vs 72.1% HR 0.75, p 0.03
GOG-0218 Incompletely resected Stage III or IV ovarian cancer, Carbo AUC 6 Paclitaxel 175 mg/m2 q 3 weeks cycle 1-6. Placebo vs Bevacizumab15 mg/kg q 3 weeks from cycle 2 -6. Bevacizumab 15 mg/kg q 3 weeks from cycle 2 -22. Primary endpoint = PFS	Median progression-free survival was 10.3 months in the control group, 11.2 in the bevacizumab-initiation group, and 14.1 in the bevacizumab-throughout group 4 month PFS advantage
ICON 7 suboptimally cytoreduced cancer carboplatin (area under the curve, 5 or 6) and paclitaxel (175 mg per square meter of body-surface area), given every 3 weeks for 6 cycles, or to this regimen plus bevacizumab (7.5 mg per kilogram of body weight), given concurrently every 3 weeks for 5 or 6 cycles and continued for 12 additional cycles or until progression of disease	Progression-free survival at 36 months was 20.3 months with standard therapy, as compared with 21.8 months with standard therapy plus bevacizumab (hazard ratio, 0.81, P=0.004

Intraperitoneal chemotherapy for optimally debulked cancer, **Day 1.** IV paclitaxel (135 mg/m^2) over 24 hours on day 1, **Day 2** IP cisplatin (100 mg/m^2), **Day 8** IP paclitaxel (60 mg/m^2) for six cycles every 21 days 42% patients could complete six cycles due to toxicity vs IV paclitaxel (135 mg/m2/24 h) followed by IV cisplatin (75 mg/m2) on day 2	Median overall survival of 65.6 months for the IP arm vs 49 months for the IV group

Second line chemotherapy for platinum sensitive epithelial ovarian cancer

- 6 x Carboplatin AUC 5 IV push and liposomal doxorubicin 30 mg/m^2 IV over 30 min q 28 days
- 6 x Paclitaxel 175 mg/m^2 IV over 3 h and carboplatin AUC 5 IV over 1 h q 21 days
- 6 x Paclitaxel 80mg/m^2 IV over 1 h weekly days 1, 8, and 15 and carboplatin AUC 6 IV over 1 h on day 1 q 21 days
- 6 x Docetaxel 75 mg/m^2 IV over 1 h and carboplatin AUC 5 IV over 1 h q 21 d
- 6 x Gemcitabine 1000 mg/m^2 IV over 30 min on days 1 and 8 and carboplatin AUC 4 IV over 1 h on day 1 q 21 days

Second line chemotherapy for platinum resistant ovarian cancer

Drug	Response rate	SD	n
Liposomal doxorubicin	20%		273 compared with topotecan
Weekly paclitaxel (80 mg/m^2 weekly for three weeks followed by one week off)	21%		48
Docetaxel (75 to 100 mg/m^2 given every 21 days)	23%		32
NAb-paclitaxel (100 mg/m^2 given weekly for three weeks on/one week off)	23%	36%	51
Oral etoposide (50 mg/m^2 daily for 21 days every four weeks	27%	-	41
Gemcitabine (1000 mg/m^2 on days 1 and 8 of a 21-day cycle)	9%		195 compared with liposomal doxorubicin
Pemetrexed (900 mg/m^2 every 21 days)	21%		51
Topotecan (1.25 mg/m^2/day on days 1 through 5 of a 21 day cycle)	19%		192 two different topotecan regimens compared

SD = stable disease n= number of patients in study

AURELIA study chemo +/- bevacizumab in platinum resistance ovarian cancer		
AURELIA chemotherapy with or without bevacizumab	Median PFS 3.4 months with chemo versus 6.7 months with Bev + chemo.	
	Response rate	Number of patients
Paclitaxel 80 mg/m^2 days 1, 8, 15, and 22 every four weeks + bevacizumab	52%	115
Topotecan 4 mg/m^2 days 1, 8, and 15 every four weeks (or 1.25 mg/m^2 on days 1 through 5 every three weeks) + bevacizumab	23%	120
Pegylated liposomal doxorubicin (PLD) 40 mg/m^2 day 1 every four weeks + bevacizumab	18%	126

Few of Current Clinical Trials in UK

ICON 8 – G126
An International Phase III Randomized Trial Of Dose- Fractionated Chemotherapy Compared To Standard Three- Weekly Chemotherapy Following Immediate Primary Surgery Or As Part Of Delayed Primary Surgery For Women With Newly Diagnosed Epithelial Ovarian Fallopian Tube Or Primary Peritoneal Cancer.

- **Arm 1**

6 cycles of therapy administered on day 1 every 3 weeks consisting of:
Paclitaxel 175mg/m2 administered over 3 hours
Carboplatin AUC5 administered over 30 minutes to 1 hour

- **Arm 2 (dose-fractionated paclitaxel)**

6 cycles of therapy each administered over 3 weeks consisting of:
Paclitaxel 80mg/m2 administered over 1 hour on days 1, 8 and 15
Carboplatin AUC5 administered over 30 minutes to 1 hour on day 1

- **Arm 3 (dose-fractionated carboplatin and paclitaxel)**

6 cycles of therapy each administered over 3 weeks consisting of:
Paclitaxel 80mg/m2 administered over 1 hour on days 1, 8 and 15
Carboplatin AUC2 administered over 30 minutes on days 1, 8 and 15

AUC 5 by Wright formula, AUC 6 by Cockroft and Gault/ Jeliffe method of GFR calculation.
AUC 2 by any method acceptable

Must commence chemotherapy within 8 weeks of operation.

Post-operatively, chemotherapy should recommence between 1- 6 weeks after delayed primary surgery. To maintain treatment intensity, it is recommended that chemotherapy should be restarted as soon as the patient is fit enough, therefore it should ideally recommence within 1-2 weeks following surgery. However, there should be at least 1 week between DPS and post-operative chemotherapy to allow for post-surgical recovery.

Ariel 2
Single agent phase II study PARP inhibitor rucaparib in platinum sensitive relapsed high grade serous and high grade endometrioid cancer in patients without BRCA 1 or 2 mutations

Ariel 3
Relapsed high grade serous ovarian cancer that has responded to 2nd, 3rd, 4th line platinum. Placebo vs rucaparib.
A Phase 2, Open-Label Study of Rucaparib in Patients with Platinum-Sensitive, Relapsed, High-Grade Epithelial Ovarian, Fallopian Tube, or Primary Peritoneal Cancer

Rucaparib 600mg orally twice daily.
Treatment is continuous in 28-day cycles until disease progression or the patient withdraws for any other reason.

BritROC.
BriTROC is a national ovarian cancer tissue and biospecimen bank. Tumour biopsy taken at the time of tumour recurrence is compared to the initial biopsy to study genetic changes between the two samples to understand chemotherapy resistance.

.Eligibility - relapsed high grade serous cancer amenable to biopsy.

OCTAVE
Platinum resistant ovarian cancer phase 1 intraperitoneal adenovirus.

Metro-BIBF
Randomised phase II
Oral cyclophosphamide with or without ninetedinib.

ANZ GOG – 0701
Looking at whether Palliative Chemotherapy Improves Symptoms in Women with Recurrent Ovarian Cancer.
Non interventional study.
Eligibility: Platinum resistant/refractory epithelial ovarian cancer with a life expectancy greater than 3 months, who are about to start palliative chemotherapy.
Health related qualify of life (HRQL) questionnaires prior to receiving each cycle of palliative chemotherapy.
Ref: http://www.anzgog.org.au/uploads/ANZGOG%20Trial%20-%20Symptom%20Benefit.pdf

www.ingramcontent.com/pod-product-compliance
Lightning Source LLC
Chambersburg PA
CBHW072308170526
45158CB00003BA/1244